Robert Koch (1843-1910)

German doctor, discovered the different kinds of bacteria that caused certain diseases, such as tuberculosis.

Sir Howard W. Florey (1898-)

Englis
first t
a hum
helpe
lives i

Sir Alexander Fleming (1881-1955)

English doctor, discovered penicillin in 1928. His experiments in developing the wonder drug earned him the Nobel Prize.

Ernest B. Chain (1906-)

Florey's co-worker, was the first to produce pure penicillin. He shared the 1945 Nobel Prize with Fleming and Florey.

This book was edited under the supervision of Nancy Larrick, Ed. D.

Junior Science Books are dedicated to all children who are eager to know more about nature and the world they live in. Written especially for young readers, each Junior Science Book has been carefully tested by the Spache Readability Formula. These books can be read by primary grade children and enjoyed by young readers through the elementary grades.

Junior Science Book of

BACTERIA

By

Gerald S. Lietz, M.D.

Illustrated by

Stephen Peck

JUNIOR SCIENCE BOOKS

GARRARD PUBLISHING COMPANY

CHAMPAIGN, ILLINOIS

The author and editor are grateful to Dr. G. P. Gladstone, Reader in Bacteriology, University of Oxford, for reading the manuscript of this book and advising on scientific content.

 Published in Champaign, Illinois, by Garrard Publishing Company.

Manufactured in the United States of America

Library of Congress Catalog Card Number: 64-10067

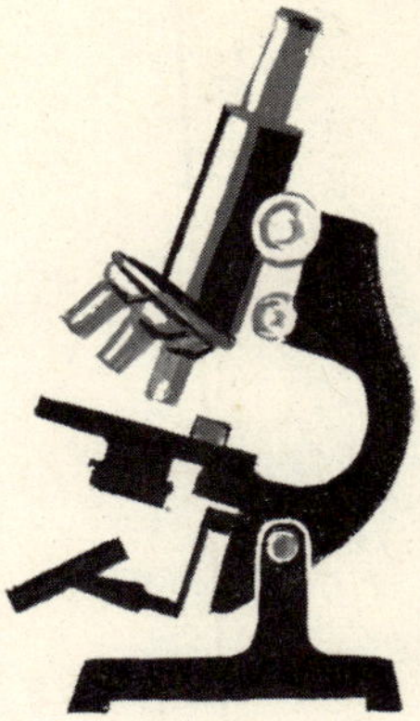

Contents

Wee Beasties

Almost 300 years ago a Dutch draper looked at a drop of old rainwater through a strong magnifying glass. Anthony Van Leeuwenhoek (pronounced *Lav*-un-hok) had made the magnifying glass himself, perhaps to check the cloth in his shop. What he saw made history. Tiny objects were moving in the water. Van Leeuwenhoek thought they were little animals. And so they became known as "wee beasties."

But the amateur scientist made one mistake. He didn't see animals. He saw bacteria, which are simple, one-celled plants.

Van Leeuwenhoek discovered bacteria in the year 1676. He could see only the larger bacteria with his magnifying glass. Today scientists can see the smallest kinds with powerful microscopes. Some bacteria are so tiny that several thousand could rest on the period at the end of this sentence.

We usually think of bacteria as germs. But germs are only one kind of bacteria. Germs can cause illness. One germ causes measles. Another kind brings mumps. Still other germs are responsible for many other diseases.

But very few bacteria are harmful. In fact, bacteria are man's best friends. We could not live without them. We would starve.

Bacteria in the soil make it possible for plants and trees to grow. The soil bacteria turn dead plants and dead animals into humus. Humus is the rich, dark topsoil. It has as much food as a supermarket—food for plants, that is.

A modern microscope

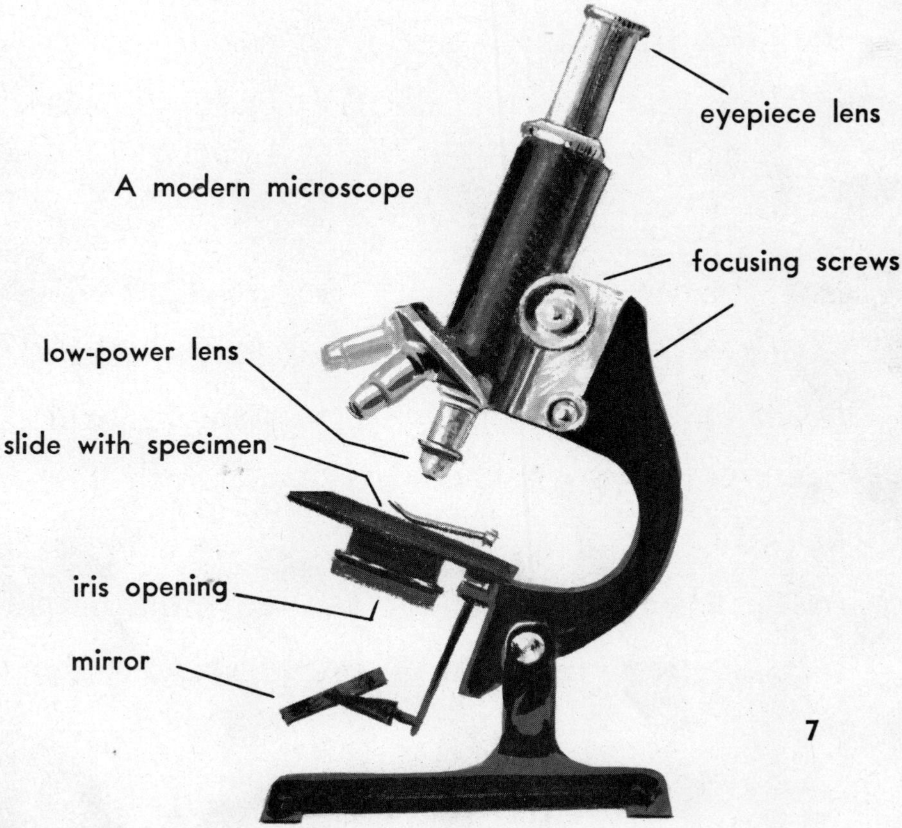

All kinds of vegetables and grains and fruit trees get food from humus. All kinds of animals get their food from vegetables and grains and fruits—or from other animals. So life depends on soil bacteria.

Without bacteria, a dead tree or plant wouldn't rot. It would just stay on top of the ground. It would never become part of the soil. Nothing else could grow in its place.

There are many other kinds of good bacteria. Cottage cheese is made with the help of bacteria. The dairy puts special bacteria into milk. They change milk into cottage cheese.

All kinds of cheese are made with bacteria. Different bacteria are used for each kind of cheese.

Other helpful bacteria turn grape juice into wine. Still others help to change corn, barley and other grains into alcohol.

Bacteria help in many other ways. Scientists are finding new ways to use the good ones. They are also finding more ways to fight the bad ones.

Bacteria Are Everywhere

No matter where you go, you are always surrounded by bacteria. The ground is full of bacteria. The air is full of them. They are on everything you see.

Touch anything and you will touch bacteria, too. They are on your clothes. They are all over this book. Many are in your stomach and lungs. When you eat and breathe, bacteria get inside your body.

Most bacteria live in the ground. A pound of dirt has hundreds of millions of them. Because they are so tiny, they are easy to move. Every little breeze picks up bacteria and carries them away.

When it rains, bacteria are washed out of the air. They are washed from plants and from dirt. Running water from streets and fields is full of bacteria. The water runs into streams, the streams empty into rivers. Rivers carry bacteria for many miles.

You can't see bacteria travel, but you can see how they travel. Watch the wind stir up some dry leaves and blow them around. Watch the wind lift a scrap of paper higher than a house. The leaves and the paper carry bacteria. There are many more bacteria floating free in the breeze.

Birds carry bacteria for long distances. Airplanes, trains and ships bring bacteria from other countries (and take our bacteria abroad). Dogs, cats, horses, cows and all kinds of wild animals carry bacteria on their bodies. When a cat rubs against a

chair, he rubs off some of his bacteria. He picks up some others from the chair.

People are great movers of bacteria. Your friends bring you bacteria from their houses, and you carry bacteria to theirs.

You pick up bacteria from the person sitting next to you on a bus. You get some from the letter the postman hands you. When you cough, you scatter bacteria into the air.

Suppose you live in New York, or New Orleans or San Francisco—or any other port city. You could bring home bacteria from Africa. How? Perhaps the bacteria were on bags of cocoa beans shipped from Ghana. A truck driver had hauled a load of beans to a chocolate factory. Some of the bacteria had stuck to his clothes. When his work was done, he rode home on a bus. You sat next to him. You picked up some of his African bacteria.

The chances are that the foreign bacteria

Bags of cocoa beans are loaded in Ghana and shipped to the United States.

won't be much different from those that grow in the United States. Bacteria pay no attention to boundary lines.

These microscopic plants can live where people would soon die. They aren't killed by the cold of the polar regions. They can stand the heat of the desert. But no bacteria are fireproof. All would die if dropped into a good bonfire.

But most bacteria die without being burned up. Many run out of food and starve. Others die from old age.

You might pick up African bacteria from the truck driver who hauled the beans to the factory.

How Bacteria Live

Green is a color that means a miracle. Nearly every plant in the world has green leaves. The green comes from *chlorophyll.* And chlorophyll makes the food the plant needs.

Food for plants is sugar and starch. When sunlight falls on a leaf, the chlorophyll goes to work. It takes a gas called carbon dioxide from the air. It takes water from the ground. From gas and water the chlorophyll makes sugar and starch.

Every blade of grass is a tiny food factory when the sun shines. Scientists still don't know exactly how the food is made. They know that the plant must have green coloring. It must also get sunlight.

Bacteria are not like other plants. They are much smaller. They do not have roots, stems, leaves or flowers. And they generally do not have green coloring. This means they cannot make their own food. They have to find food that is already made.

Bacteria can feed on almost anything. They eat every kind of food used by man and animals. They eat a lot of things that don't seem tasty at all—such as wood, paper, paint, tar and glue.

Bacteria have different shapes. But all of them have a one-piece body. This one piece of life is called a cell. The human body has billions and billions of cells.

A single bacterium is a simple little plant. It is built something like a grape. It has an outer coat called a cell wall. Inside the wall is a clear, jelly-like body. The best way to see this body under the microscope is by coloring it. Scientists use colored dyes to stain bacteria.

The jelly-like body holds food for the cell. The food is taken in through the cell wall.

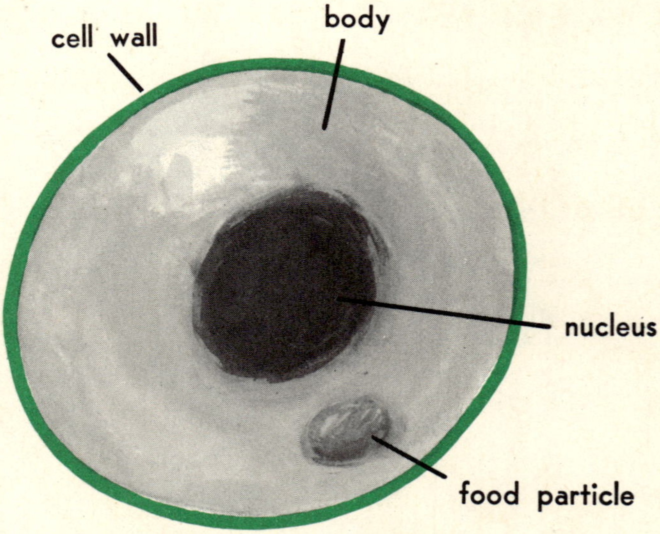

A simple plant cell looks like this under a high-powered microscope.

Some bacteria have tiny hairs. When these hairs wiggle, the cell moves much as a tadpole moves through water.

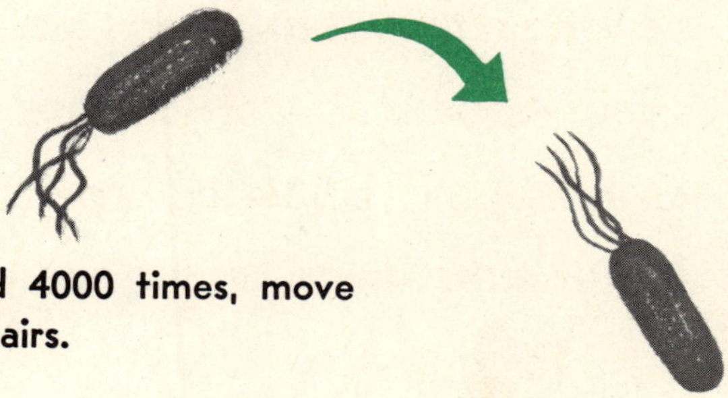

The bacteria, magnified 4000 times, move with the help of tiny hairs.

Some bacteria are round, like marbles. They may grow in a chain or in a bunch, like grapes. All of the round ones are called *coccus* (*cock*-us). One of these—the streptococcus—can cause a sore throat.

Chains of streptococci, the bacteria that causes sore throat, magnified 3000 times.

Another family of bacteria is rod-shaped. Through the microscope they look like paper clips. Sometimes they are single. Others grow in pairs or in a chain. The rod-shaped bacteria are called *bacillus* (ba-*sill*-us). Bacillus germs cause pneumonia and tuberculosis.

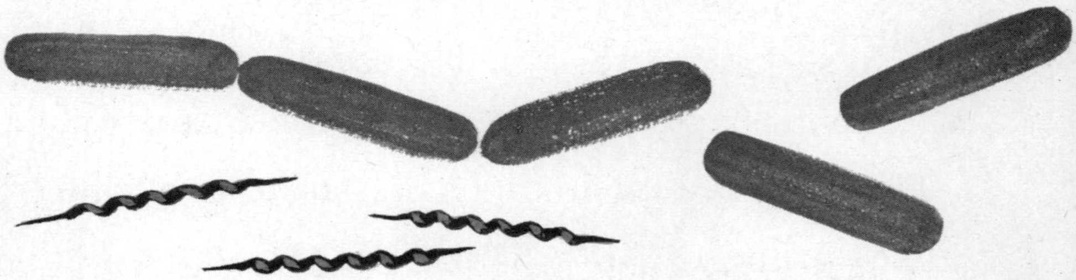

Rod-shaped bacillus and spiral-shaped spirochetes, greatly magnified.

A third kind of bacteria is spiral, like a corkscrew. These are the *spirochetes*. (*spy*-ro-keets).

Altogether there are about 150,000 different bacteria in these three shapes.

New bacteria are easily made by old bacteria splitting in two. A cell squeezes itself in the middle, as if a belt were being pulled tight. When the two halves fall apart, each is a separate cell. Each of the two cells can soon do the same, and then there are four. Next the four cells become eight, and the eight become sixteen.

You can see some big numbers ahead—and they are really big. One bacterium could become 66 billion in only 12 hours.

Why don't bacteria take over the whole earth? What stops them? Luckily, most bacteria starve to death. Others dry up.

A cell squeezes itself in the middle. The halves fall apart and become separate cells.

Watch a rotten spot on an apple. Bacteria are at work. They make the brown spot bigger and bigger until the whole apple is rotten. A small colony of bacteria has grown into billions and billions. But the rotten apple will finally dry up. There will be no more moisture and food. The bacteria will be too crowded to keep multiplying. They will stop growing, and most will die.

Not all bacteria die of starvation or dryness. Some just shrink, like a grape drying into a raisin. The cell walls get hard. The bacteria won't divide any more,

Until food and moisture are used up, billions of bacteria can live in a rotten apple.

but they are still alive. If they find food and moisture, they will grow again. They are like the dried seeds of corn or wheat.

You will remember that each bacterium is only one cell. How can a plant with only one cell take in food? It's done in an interesting way. Juices from the jelly-like body seep out through the cell walls. These juices are powerful. They can turn milk sour; they can make wood rot. They digest food outside of the cell. Then, when the food is soft enough, it comes through the wall and feeds the cell.

The smallest bacteria are called viruses. They can be seen only with the electron microscope, which makes things look 25,000 times larger than they are. Viruses are some of our oldest enemies. They cause many diseases, including mumps, chicken pox, measles and the common cold.

Fungi

Fungi (fun-ji) are not true bacteria, but they are much the same. They live and grow wherever bacteria grow. When bacteria are grown in the laboratory, fungi often grow with them. In fact, these plants are very much like Mary's little lamb. They follow where they are not wanted.

To know bacteria, you must know about fungi.

Unlike bacteria, many kinds of fungi can be seen without a microscope. In woods and fields you may have noticed mushrooms of many colors—white, gray, yellow, orange, red, brown and purple. All mushrooms are fungi.

If you have ever stepped on a puffball, you know another kind of fungi. And the shelf-like growths on the sides of tree trunks and rotting logs are also fungi.

Two kinds of fungi: puffball (left) and mushrooms (right).

But most fungi are as tiny as bacteria and can be seen only with the microscope. Like bacteria, they have no green coloring. They cannot make their own food, so they live on dead or living plants and animals.

There are so many different fungi that scientists have not been able to count them. There may be 75,000 kinds—or even more.

Among the common kinds of fungi are the molds and yeasts. Molds grow on bread, bacon and other foods. Molds also appear on the skin of spoiling fruit and on jellies and jams. Leather shoes left in a damp place will get a greenish mold.

Molds and other fungi grow from a tiny bud or *spore*. These spores are carried through the air as bacteria are. When a spore falls on a moist slice of bread, it begins to swell. It sends tiny, hair-like roots into the bread.

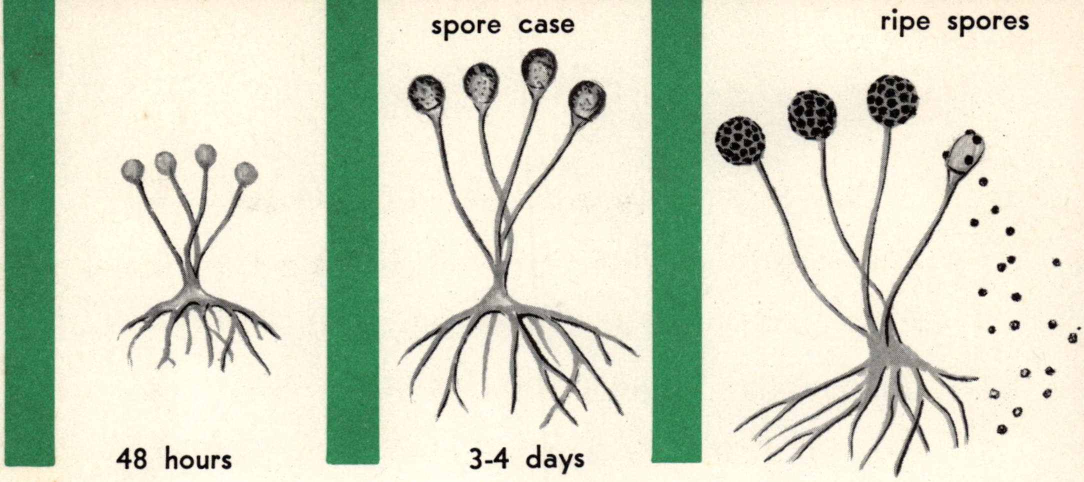

How bread mold grows, magnified 200 times.

Soon these threads make a web you can easily see. More threads grow up from the food. On these a tiny black bud will grow, no bigger than the head of a pin. When the bud gets ripe, it bursts. More than 50,000 black spores are set free. Many of them will be carried by the air to near or far places. If they land on damp food, they will start new colonies of mold.

Molds get their color from the millions and millions of spores. Some are black, some are brown, some are green, some are pink.

Fungi that grow on dead trees and plants are good friends of mankind. Like the bacteria, they are trash removers and soil builders. When a tree dies, fungi and bacteria start colonies in the roots, the trunk and the branches. The whole tree begins to rot. And the soil is richer because of the rotten wood.

Leaves and soft stems of weeds, grass and vegetables rot much faster than wood. The bacteria and fungi get rid of them in one winter.

Fungi that grow on dead things are called *saprophytes* (*sap*-ro-fites). Fungi that grow on living plants and animals are called *parasites* (*pair*-a-sites). As you might guess, the parasitic fungi have caused plenty of trouble.

Between 1845 and 1851 about 750,000 people died in Ireland because of a fungus.

This fungus made the potato vines rot. Potatoes were the main food of the Irish people. When the crops failed, men, women and children died of starvation.

Modern farmers are always watching for wheat rust, a mold that attacks the stems. Scientists have been able to breed new kinds of wheat that resist the rust. But then new kinds of mold appear, and the scientists must start over again.

One fungus makes purplish spikes on the heads of rye. A drug called *ergot* is made from this fungus. Ergot used in small doses stops bleeding. But too much of this fungus is dangerous. People have been poisoned by it. As we shall see later, there are some fungi of great value. They are made into drugs which kill germs.

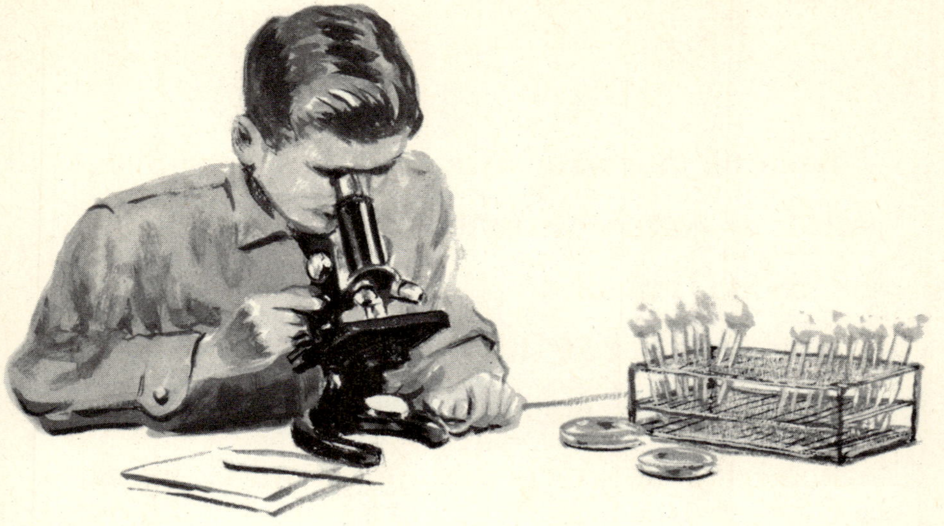

Growing Bacteria

Scientists from all over the world are constantly studying bacteria. They seek new ways of using good bacteria. They want to find better ways of fighting harmful bacteria.

Billions and billions of bacteria are grown in thousands of laboratories. Many kinds of experiments are tried to see how the bacteria behave. Sometimes important discoveries are made.

Bacteria are grown in a small glass dish, called a *culture dish.* It looks much like the round top of a plastic freezer dish. It has a cover that fits snugly to keep out other kinds of bacteria.

Foods for bacteria are meat juice, sugar and vitamins. Bacteria often grow best when these foods are served in the form of a jelly. The jelly is made from the juice of seaweed and is called *agar.* Foods are mixed with the agar when it is hot. The mixture becomes a firm jelly when it cools.

The agar in the culture dish becomes a firm jelly.

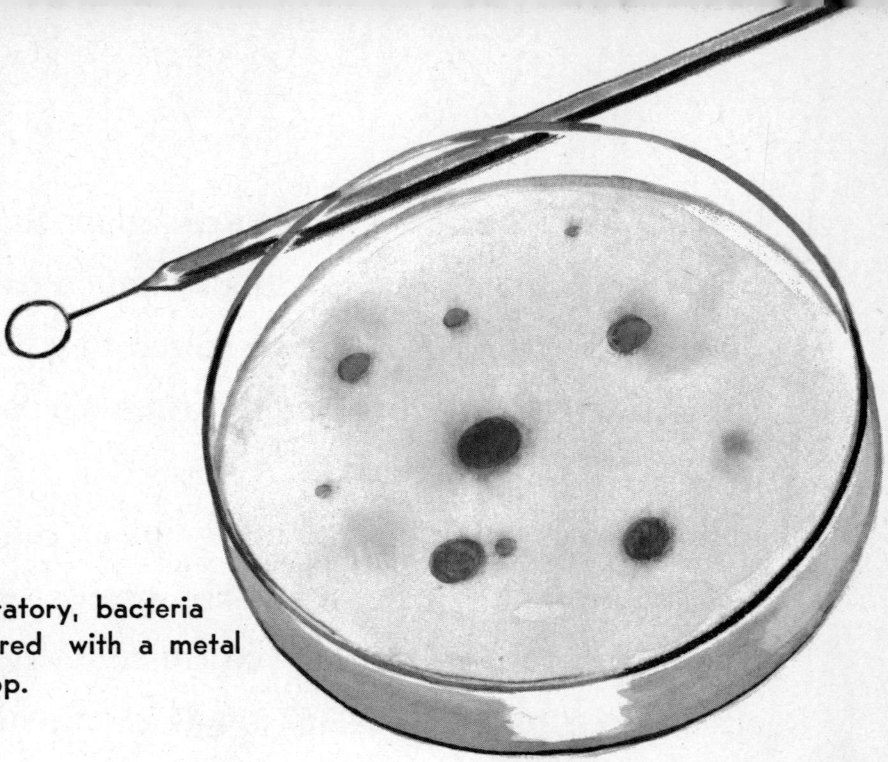

In the laboratory, bacteria are transferred with a metal bacteria loop.

The laboratory probably has some of the bacteria it wants to grow. If not, they can be had from another laboratory. The scientist touches these bacteria with a metal rod like a pencil. Then he touches the agar with the rod. The rod carries bacteria to the agar. Next the new culture dish is put into an oven just warm enough for bacteria to grow. The oven is called an *incubator*.

In a few hours a small spot will show on the agar. The spot often looks like a dewdrop. Actually it is a colony of bacteria. It gets bigger as the bacteria keep dividing. Soon the colony will have billions of bacteria.

It takes a microscope to see one or two bacteria. A tiny piece of a colony is put on a piece of glass called a *slide*. A drop of dye is added to make the cells stand out. Then the scientist can look into the microscope and see cells 1,000 times larger than they really are.

If the scientist uses the electron microscope, the cells will look 25,000 times larger. Imagine a Jefferson nickel 25,000 times larger. Thomas Jefferson's nose would be 200 feet long and rather noisy to blow.

Molds can be grown in the same way as bacteria. A small amount of the mold or some of its spores is planted on agar. Soon a colony will grow.

You can grow bacteria and molds, too. You can grow a mold very easily on bread. Wet a piece of fresh white bread with a little water. Put it on a piece of aluminum foil. Let it stand in a warm room for half an hour. Then cover the bread with a dish or cup and leave it in a warm place.

To grow mold, cover moist bread and leave it in a warm place.

Each day look at the bread with a magnifying glass. On the second day you will see tiny white hairs growing on the bread. On the third day there will be many more.

By the fourth day some of the white hairs will turn black at their tips. The magnifying glass will show you tiny black balls. Each ball has thousands of black spores. In another day the bread will be covered with a black mold.

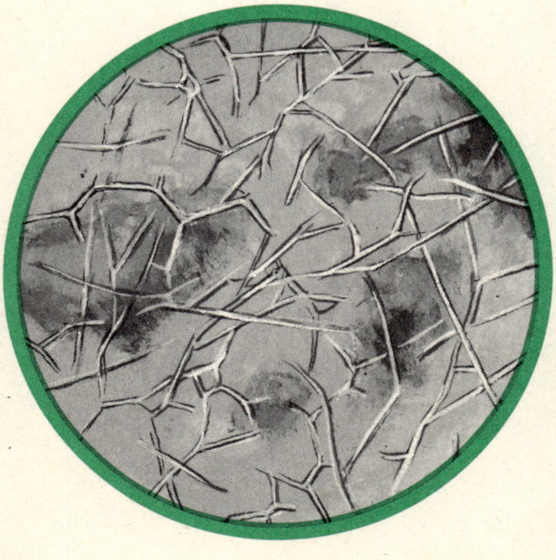

top

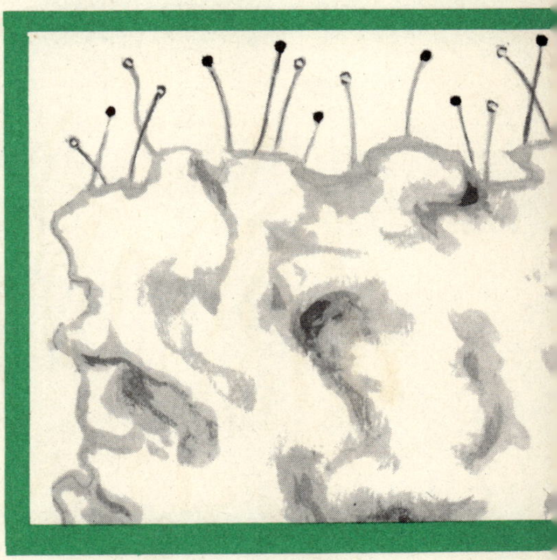

side

Your bread mold will look like this through a magnifying glass.

This mold began when the bread was left uncovered. Black spores floating in the air settled on the bread and began to grow.

You can use these spores to start another mold, as a scientist does. Wet another piece of bread and put it on aluminum foil. Take a toothpick and lift some black spores from the mold. Touch these spores to several places on the fresh bread. Quickly cover the bread so spores from the air won't settle on it.

Watch the new culture each day. The new mold will grow just as the first one did.

The black mold likes the starches found in bread. A blue-green mold likes the sugars of fruit. You can grow this mold, too. You will need a small piece of Roquefort cheese and a slice of lemon or orange.

Put the fruit slice on a piece of foil. With a toothpick take a speck of mold from the cheese. Touch this to the fruit where it meets the rind. Cover quickly and put the culture in a warm, dark place.

Again you will see the cobwebby threads growing. By the fourth day, the white colony will change color. This time you have grown a blue-green mold.

When Germs Attack

Most bacteria are harmless. But germs are enemies. If they begin to live and grow in our bodies, we get sick. We have what a doctor calls an *infection*.

Germs get inside of us every day. We breathe them through our noses. We eat them with food. Sometimes we get them into the blood when our skin is cut.

Generally the body can kill the invading germs. But not always.

Each germ goes to the part of the body where it can grow best. The germs of mumps settle in the saliva glands in front of the ears. Bacteria that cause colds like the nose and throat.

All germs need food. They get it by eating the cells of our bodies. Germs also need a pleasant climate. The temperature inside us is just right—as warm as a warm summer day.

Food and warmth make germs multiply. Soon there may be millions inside us.

As some germs grow, they make a poison. It is called a *toxin.* This toxin may be as poisonous as rattlesnake venom. It kills our body cells where it attacks. Many times the toxin spreads to other parts of the body.

Other germs make no toxins. For example, the cold germ doesn't kill our body cells. It simply makes our nose and eyes burn.

Luckily, our bodies fight back when germs invade. We are defended by the *white blood cells.* These cells are born in our bones. They become part of the blood, and they go wherever it goes. They can kill germs. When more germs attack, more white blood cells are made.

The body protects itself in other ways. The skin is an armor to keep germs out. The acid of the stomach kills germs we swallow. The nose traps many germs and keeps them from getting into our lungs.

But our defenses don't always work. Sometimes the germs make headway. We get sick. We have an infection. Then our body tries something else: a fever. Our

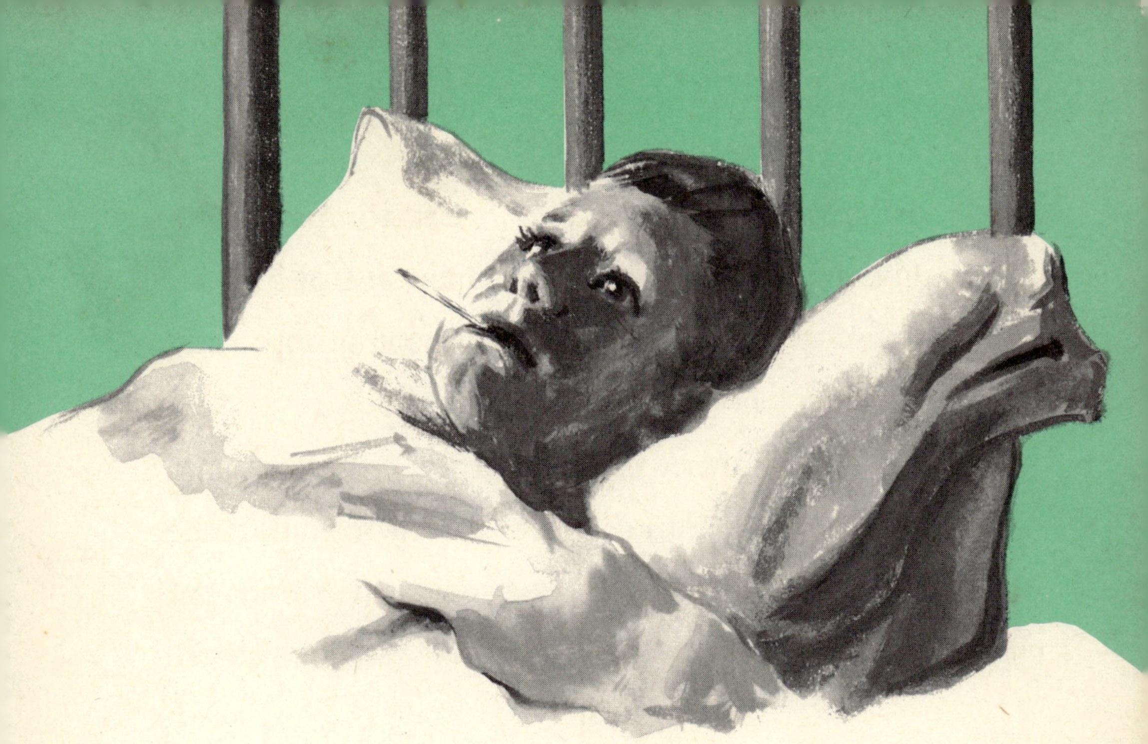

temperature goes up from a normal 98.6 degrees to 100, 101, 102 or even higher. The extra heat makes the germs weak and lazy. Then they are easy to kill.

Infections may also be in the skin. A boil is an example. The invading bacteria attack the skin. White blood cells are brought in to fight them. Some of the skin cells are destroyed in the battle. The skin becomes red, hot and swollen.

If the body wins, the infection will soon be over. The skin will heal. But if the germs are stronger, more and more skin will turn red and sore. The body will then need help from outside.

Help can come from the wonder drugs. They have saved thousands and thousands of lives by killing bacteria.

The Wonder Drugs

You have probably seen a rotten cantaloupe. The skin was covered with a green mold. From this particular mold the first wonder drug was made. The drug is called *penicillin*. It kills many germs that can kill people.

Germs are one kind of bacteria. Infections caused by germs used to kill thousands of people *each day* all over the world. Pneumonia was one of the worst infections. In the United States alone, pneumonia took 200,000 lives every year.

Many other people died from scarlet fever, meningitis and blood poisoning.

For hundreds of years men sought a medicine to destroy germs in the body. The medicine had to be safe. It had to kill bacteria without harming the patient. Many drugs were tried. Some knocked out the germs but made the patient sick.

Long ago the Chinese used a soybean mold to treat sores on the skin. They learned that moldy leather shoes helped to cure foot infections. But nothing was found to destroy infection inside the body.

In 1928 Dr. Alexander Fleming was working in the basement laboratory of St. Mary's Hospital in London. Dr. Fleming was the hospital's bacteriologist. He noticed that a plate of deadly cocci bacteria had been ruined by a stray mold. The mold spores had drifted into the laboratory with the air.

An idea flashed through Dr. Fleming's mind. If this mold could kill cocci bacteria, it might be valuable. So he saved it and grew some more. The penicillin made by the mold killed many kinds of germs. And it seemed safe. The drug was tried on a rabbit. The rabbit wasn't harmed at all.

The trouble was that they could not make much penicillin in the laboratory. Nobody knew how to make large quantities of it.

Ten years passed. At Oxford University, Dr. Howard Florey was hunting for a drug to use against infections. He remembered a report that Dr. Fleming had written on his experiments. Dr. Florey came to London and went to the hospital laboratory. Dr. Fleming was still there—and so was the mold. For ten years Dr. Fleming had kept it growing.

Dr. Fleming (right) shows Dr. Florey the penicillin mold.

More penicillin was made. Finally it was tried on people who were dying from infections. The patients got well. But there was only enough penicillin to save a few lives. Making more would be very expensive. Private drug companies were not interested.

In 1941 Dr. Florey and Dr. N. G. Heatley came to the United States. They were seeking a cheap way to make plenty of penicillin. The U. S. Department of Agriculture was ready to help. So the British scientists went to the Department's Northern Regional Research Laboratory in Peoria, Illinois.

A long series of experiments began. There are many different kinds of penicillin molds. Which would be the best? One girl collected hundreds of samples of spoiled fruit and vegetables from all over

town. They were covered with different kinds of molds. A rotten melon brought in by "Moldy Mary" was the prize. This mold could be made in huge amounts by using water in which corn had been soaked. Milk and sugar were added to the water. The mold made 100 times more penicillin than Dr. Florey's mold did.

The government laboratory had found the secret of making penicillin in large quantities. Then the private drug companies were interested. All of the big firms wanted to make and sell penicillin.

Manufacturing began late in 1942. Soon penicillin was saving lives in hospitals and on battlefields. It is still the best all-purpose wonder drug. But it will not kill every kind of germ.

Penicillin is grown in fermentation silos.

A drug made from bacteria or fungi to fight other bacteria is called an *antibiotic*. The word comes from the Greek language. It means "against life." The antibiotic is against the life of harmful germs. Penicillin is an antibiotic.

Since penicillin, other wonder drugs have been found. They are made from other molds and bacteria found in the soil. The wonder drugs can cure most kinds of infection.

A doctor can choose the right drug for each infection. A drop of each different antibiotic is put on a separate spot on a culture dish. Then some germs from the infection are put next to each drug. If the germs grow, the drug next to them won't work. If the germs stop growing, that drug can be used.

Another use for antibiotics was discovered by accident. Farm animals were being fed with the mash left over in making antibiotics. This mash contains Vitamin B_{12} which makes animals grow faster. Chickens and pigs and lambs and calves can be sent to market sooner. The farmer doesn't have to buy so much feed. He can make more money.

Sometimes the mash had a little of the antibiotics. When this happened, the animals grew still faster. Scientists have not been able to find out exactly why. But the evidence is clear: Antibiotics speed up the growth of young animals.

Vitamin B_{12} makes animals grow faster. The pigs are the same age, but the larger was fed a B_{12} diet.

Armor Against Infection

Once you have been sick with chicken pox you will never have it again. Your body has made an armor against chicken pox germs. This armor is called *immunity.*

The diphtheria germ makes a poison, or toxin. Your body makes an *antitoxin* to fight this poison. How this is done no one knows. It is one of the wonderful things the body can do.

Other germs are handled in different ways. The armor against the germs of chicken pox, measles and smallpox is a material like an antitoxin. But it does not fight the poison. It simply holds on to the germs and keeps them from attacking your body cells. While this material is attached to the germs, your white blood cells eat up the invaders.

Each time you are sick a different protective material or antitoxin is made. It is made especially for each germ. It stays in the blood. It may protect you for years and years. But you are protected against only the sickness you have had. For example, the chicken pox armor won't protect you from the measles.

You can also get immunity by being vaccinated. You were probably vaccinated for smallpox before you were one year old.

A doctor scratched the skin on your leg or arm. A drop of smallpox virus was rubbed into the scratch. In a few days the skin turned red at that spot. Your body was building protection against smallpox.

Where did the smallpox germs come from? They were taken from a healthy young calf. The calf had been vaccinated, too, with the same virus.

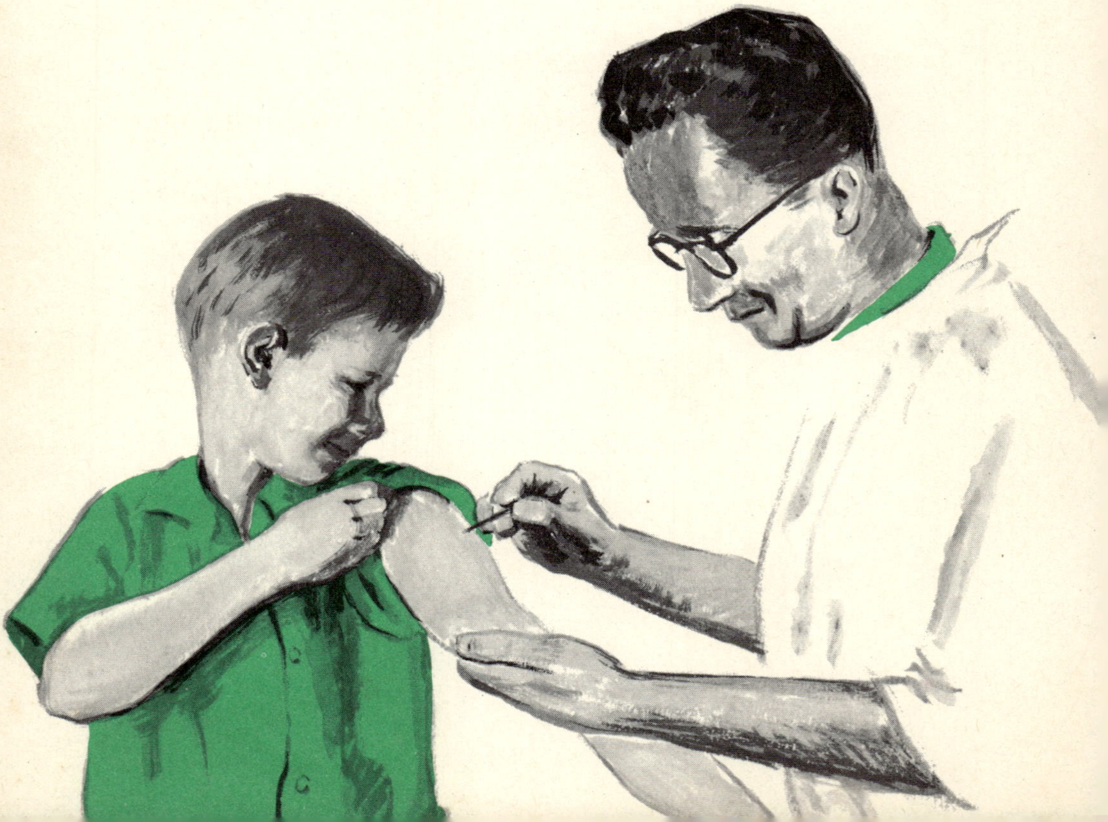

But why don't you get smallpox when you are vaccinated with smallpox germs? Because the smallpox germs in the calf had lost a lot of their power. Your body can overcome the germs by making the protective material.

Most young children are also vaccinated against diphtheria, tetanus, polio and whooping cough. Their bodies build antitoxins or protective material for those germs. It's the same as getting the disease —except much pleasanter.

Much of the antitoxins and protective material is lost as the years pass. To keep your armor strong, you need booster shots every few years.

Controlling Bacteria

Food left in the open air will soon spoil. If the air is quite warm, the food will rot or turn sour faster.

Spoiling is caused by the bacteria that are always floating in the air. Some settle on the food and begin to eat. The more they eat, the more they grow. The more they grow, the more they eat.

Bacteria like warmth. They grow faster and eat more when the temperature is up (but not too high).

Milk is a good food for bacteria. At room temperature milk will sour in two days. But in a refrigerator milk will keep for nearly a week. Cold air slows the bacteria.

If the milk were frozen and put in the freezer, it would stay sweet for months. Meat will last nearly a year when frozen. The bacteria aren't killed by freezing. They are just knocked out. When a frozen food thaws, the bacteria get active again. They start eating and multiplying. The food will soon spoil.

High temperatures kill bacteria. Many bacteria live only a few minutes in boiling water. There are a few bacteria that cannot be destroyed this way. Their thick cell walls protect them.

Canned food is sterile and will keep for years.

Whenever all the bacteria are killed, food is said to be *sterile*. Sterile food will last for years if bacteria are kept away.

Most canned foods are cooked until they are sterile. Next the food is put in a can. The can is sealed to keep out air and bacteria. Then the can is heated enough to kill any bacteria that may have slipped in.

Canned food will keep for years. But if air gets in, the food will soon spoil.

Certain chemicals keep bacteria from growing. When these chemicals are added to food, the food will last. It will be *preserved*. One common preservative is vinegar. The liquid in a jar of olives or pickles is vinegar.

Preservatives keep bacteria from growing.

Salt has been used for thousands of years as a preservative. The American pioneers kept meat from spoiling by salting it.

Fancy chemicals aren't needed to keep our bodies from spoiling. We remove bacteria from our teeth by brushing them. We wash bacteria from our skin with soap and water.

But remember that most of the bacteria are friends. All over the earth they make food and shelter for men and animals. There would be no rich soil without bacteria. And without soil, you wouldn't be eating. You wouldn't be reading from this book, either. The paper in it came from trees that bacteria helped to grow.

Bread, Wine and Linen

Men have been using bacteria and fungi for thousands of years. They help to supply food and drink and clothing.

Nobody knows who made the first loaf of bread. It was probably an accident. The Egyptians were among the first bread makers, and we can imagine what might have happened. An Egyptian mother had made a porridge of flour and water. The flour probably came from wheat, barley and durra. Durra is a kind of grass with seeds that can be eaten.

The porridge had been left near the oven for a couple of days. (Maybe the family was tired of the same old food.) Then, strangely, the porridge began to swell up. Instead of throwing away the puffy mess, the mother put it in the oven to warm. It became light brown on top and dry inside. It was a loaf of bread.

Yeast fungi had begun to live and grow in the wet porridge. Yeast makes dough puff up. It fills the dough with thousands of air pockets. Without yeast, a slice of bread would be as flat as leather, and almost as tough.

The ancient Egyptians became skillful in making bread. But nobody learned about yeast until 5,000 years had passed. It was not until 1837 that scientists found out what yeasts are. The microscope showed that they were plants.

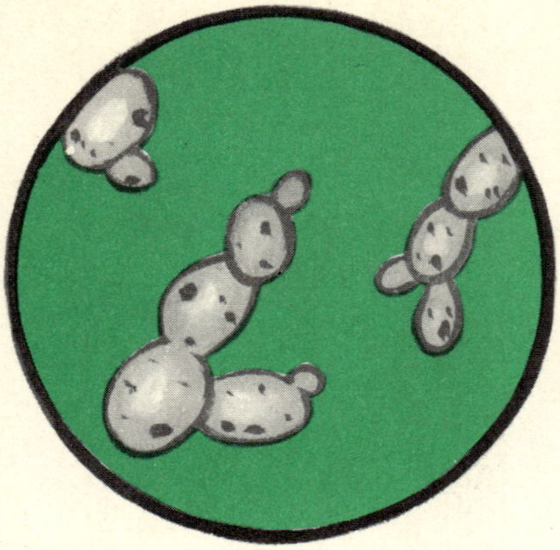

Yeast causes bread dough to rise. These yeast plants are magnified 600 times.

Bakers discovered long ago that yeast could be moved from one batch of dough to another batch. The bakers would save a little dough each day. Next day the little piece would be mixed with the fresh dough. The yeast would spread and all of the new dough would rise.

Now yeast is made in factories. It is sold to bakeries and other users. A large bakery will use more than 100 pounds of yeast each day.

The Egyptians were early wine makers, too. They knew that something turned grape juice into wine. But they did not know what made the change.

Thousands of years later scientists found the answer. There are yeasts on the skins of grapes. These yeasts make the juice ferment. When the juice ferments, it becomes wine.

You have probably used a linen handkerchief. Oddly, bacteria are needed to make linen cloth.

How? Linen is made from the stalks of a plant called flax. It is a handsome plant with bright blue flowers. But the stems are quite tough. They have to be softened and split into many fibers, like threads.

The stalks of the flax plant are strong and tough.

Workers tie the stems into bundles and put them in water about four feet deep. The bundles are covered with straw. Stones are laid on top of the straw to hold the flax under water.

In a short while bubbles rise to the surface. The flax stalks are fermenting. Bacteria are at work. In about ten days the stalks are softened. They are taken from the water and spread out to dry. Then they are split to make threads.

Linen has been used for thousands of years. The lake dwellers of Switzerland made linen in prehistoric times. The Egyptians used linen to wrap mummies.

Wherever you look, you can find bacteria and fungi at work for plants, animals and people. Without bacteria and fungi, the world would become a desert.

Index

Anthony Van Leeuwenhoek (1632-1723)

Dutch draper, is called the "Father of Bacteriology." He was the first to see bacteria, which he called "wee beasties."

Joseph Lister (1827-1912)

English doctor, was the first to use ANTISEPTICS to kill disease-carrying bacteria in hospitals. Safe operations resulted from his work.

Louis Pasteur (1822-1895)

French chemist, found that germs caused infection. He developed PASTEURIZATION to preserve food and IMMUNIZATION to control the spread of disease.

Ferdinand Cohn (1828-1898)

German botanist, was the first to classify bacteria into types. He discovered that bacteria were plants and produced spores.